TOPIC RE

The Safety of FREE STEERED VEHICLE *operations below ground in* BRITISH COAL MINES

HM INSPECTORATE OF MINES

LONDON: HMSO

©Crown copyright 1992
Applications for reproduction should be made to HMSO

Enquiries regarding this or any other HSE publications should be made to HSE Information Centres at the following addresses:

Broad Lane
Sheffield S3 7HQ
Telephone: (0742) 892345
Fax: (0742) 892333

Baynards House
1 Chepstow Place
Westbourne Grove
London W2 4TF
Telephone: (071) 221 0870
Fax: (071) 221 9178

Printed in the UK for the Health and Safety Executive 03/92

ISBN 0 11 885996 X

CONTENTS

page 1 **INTRODUCTION**

1 **FREE STEERED VEHICLE OPERATION**

1 Use of FSVs
2 Roadways in which free steered vehicles operate
3 Appointment and training of personnel
4 Risks to pedestrians from FSVs
5 Cab design and driver's vision

8 **A REVIEW OF ACCIDENTS RESULTING FROM THE USE OF FSVs**

8 Overall Accident Summary

9 **ACCIDENTS RESULTING FROM THE MOVEMENT OF VEHICLES**

9 Summary
9 Accidents to pedestrians
10 Accidents resulting from collisions
12 Clearances
12 Other accidents

13 **FSV LOAD TRANSFER ACCIDENTS**

13 Summary
13 Load out of control accidents
13 Strain and trapping injuries
13 Stumbling, falling and slipping
13 Fitting and releasing loadbinders
13 Other accidents
13 Recommendations regarding transfer

14 **ACCIDENTS ASSOCIATED WITH FSV MAINTENANCE**

14 Accident summary
14 Maintenance accidents
14 Underground garages
15 Safety awareness training

15 **OTHER ACCIDENTS INVOLVING FREE STEERED VEHICLES**

15 Accident summary
15 Roadway floors

15 **OTHER DANGERS**

15 Underground fires
16 Runaways
16 Overturning of FSVs
16 Vehicle collisions

CONTENTS
Continued

page 16   **BRAKE TESTING**

17   **DIESEL VEHICLE EXHAUST EMISSIONS**

18   **BATTERY POWERED VEHICLES USED IN COAL MINES**

19   **NOISE CONTROL**

20   **CONCLUSIONS**

20   **SUMMARY OF RECOMMENDATIONS**

23   **GLOSSARY**

INTRODUCTION

1. The introduction and development of free steered vehicle (FSV) transport in British coal mines probably represents the most significant change in underground transport for some considerable time. The purpose of this report is to examine the safety of FSV operations below ground in coal mines, and to recommend areas where safety might be improved. The report is primarily intended as an informative document for the guidance of management, supervisory staff, engineers and operatives at mines, and is based on operating and accident experience for the five-year period 1986 to 1991.

2. Self propelled, battery powered, shuttle cars were used underground in UK coal mines in the mid-1940s, and small diesel powered tractors from 1959 onwards, but their use remained limited. In the mid-1970s, because of the constraints and limitations of rope and locomotive haulage systems, the then National Coal Board reviewed transport arrangements for material handling. It was decided to examine the potential of the diesel powered load-haul-dump (LHD) machines which were used extensively abroad and in non-coal mines in this country. Between 1975 and 1981 operational trials were carried out at High Moor Colliery using suitably modified proprietary vehicles which were later developed into purpose built vehicles. The term free steered vehicle was a phrase adopted to describe the type of vehicle introduced below ground in coal mines from the early 1980s onwards and for the purpose of this report the term FSV includes shuttle cars. The numbers held increased rapidly from 188 in 1986 to 460 by 1991.

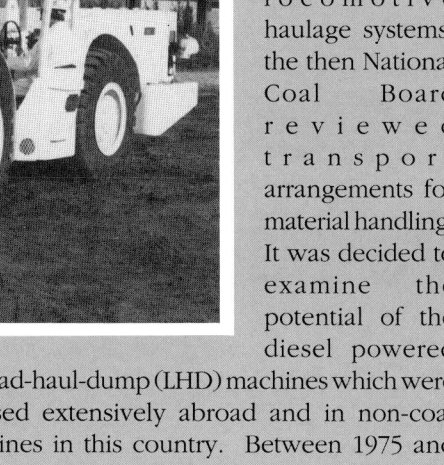

Photograph 1: A typical free steered vehicle

3. A typical FSV **(photograph 1)** is a self propelled vehicle running on rubber tyred wheels, generally operating without much roadway floor preparation, designed to be driven in either direction from a centrally placed driver's cab, and used primarily for transporting material and equipment. Pivot steering and the capacity to carry an outboard load make the FSV a most versatile and manoeuvrable vehicle.

4. The construction and use of diesel powered FSVs for use below ground in coal mines is controlled either by Local Regulations or by Consents issued under Section 83 of the Mines and Quarries Act 1954 (MQA). Both diesel powered vehicles and, by virtue of the Electricity at Work Regulations 1989, storage battery powered vehicles used in coal mines must be of a type approved by the Health and Safety Executive (HSE). It is anticipated that the present small number of battery vehicles in use will increase in the years ahead.

5. Although the report is based primarily on experience in coal mines, much of it is also pertinent to non-coal mines.

FREE STEERED VEHICLE OPERATION

Use of FSVs

6. In 1991, the number of FSVs held by British Coal Corporation (BCC) mines was 460, comprising 381 diesel powered vehicles, 16 cable reel vehicles, 13 battery powered vehicles, and 50 cable reel shuttle cars. There is a general consensus that the number of battery powered vehicles will increase over the next few years at the expense of diesel vehicles which have higher maintenance costs and provide lower availability time. The majority of FSVs were used for material transport, about a third for mineral haulage, and there were nine purpose designed personnel carriers. Two special purpose diesel powered vehicles were used for roadway floor maintenance.

7. Priority in the development of FSVs in coal mines was to transport material direct to the point of usage without intermediate transfer. The flat bed type vehicle **(photograph 2)** was developed to carry bulky material loads and heavy machinery without dismantling. FSVs fitted with loading winches became an accepted means of recovering, transferring and installing powered roof supports and other face equipment. Design has sought to provide a narrow vehicle with heavy load carrying capabilities, the most popular being 1.5m wide with a carrying capacity of around 8 tonnes, and powered by a 75kW flameproof diesel engine. Vehicles with carrying

Photograph 2: A flat bed type FSV carrying a powered support

Photographs 3a & 3b: A quick detatchable system (QDS) engaging a loading bucket

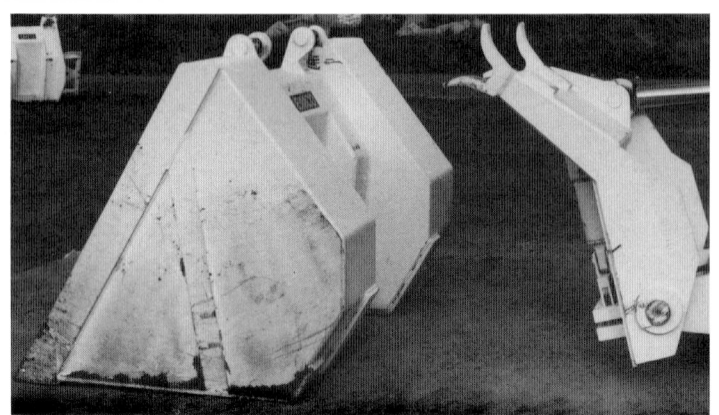

capacities of up to 20 tonnes have been developed for larger loads. A variety of vehicle attachments have been designed, such as lorry loader cranes, chock pulling arms, stone dusters, back actors, floor plough blades and roof bolting equipment.

8. LHDs, typically with two to three cubic metre capacity buckets, were introduced to assist dinting and to transport materials. Quick detachable systems (QDS) **(photographs 3a & 3b)** for interchanging buckets for material carriers were developed to make their use more versatile.

9. An early concept was to adapt material vehicles for manriding by attaching a passenger carrying 'pod' **(photograph 4)**. The conflicting requirement for soft vehicle suspension for manriding comfort against hard suspension for load carrying stability, and the limited driver vision, led to the introduction of purpose designed passenger carrying vehicles **(photograph 5)**. These 'buses' generally have a driving cab at each end to give the driver good visibility in either direction of travel, and provide seating within a protected canopy for up to 30 passengers. Recently, there has been an upsurge of interest in passenger carrying FSVs.

Roadways in which free steered vehicles operate

10. The FSV profile and roadway cross section determine operational clearances which should conform to the mandatory minimum clearances stipulated in any consent granted by HSE.

11. Collision damage to cabs or machine covers may indicate inadequate clearances. Clearance evaluation should be undertaken at the roadway design stage when the proposed function of the roadway and the intended types of FSV to be used should be determined. The evaluation must take account of any obstructions which are likely to be present in the roadway since these determine the available width for operation. Specified roadway dimensions should take into account possible reduction in cross section due to strata movement. Additional clearances should be considered for bends, junctions, material transfer stations, garages and other special places where the need is identified.

Photograph 6 shows a typical well designed FSV roadway with good clearances.

12. FSVs may continue to operate in a length of road in a mine in which clearances are reduced due to strata movement, provided safety is not impaired and steps are being taken to restore the specified clearances. The FSV Transport Rules may be amended, for instance, by introducing a slower vehicle speed.

13. The quality of the roadway floor plays an important part in the safe and successful use of FSVs in mines. Most roadways in British mines have compacted dry natural floors, but at places subjected to heavy tyre loading, such as junctions and roadways with wet floor conditions, concrete or similar surfacing may prove necessary. Where clearances are limited to the extent that vehicles cannot vary their track, rutting will inevitably occur on weak floors. This is generally made worse by the presence of water and may result in tyre adhesion being lost. Uneven floors cause vehicle tilt and bounce which reduce clearances and may affect vehicle stability. The typical FSV is far heavier than the load it normally carries and even empty vehicles cause floor wear. FSV roadway surfaces, like rail track, require positive and regular maintenance. The once held belief in the ability of LHD machines to fully maintain roadway floors has not proved true in practice. At surface opencast coal sites, road surfaces are repaired by planing off the surface and redistributing and compacting the planed material using purpose built vehicles. Similar types of machines are now on field trials in some mines **(photograph 7)**. Good mine roadway floors are a prerequisite for comfortable passenger transport.

14. A number of FSV roadways also contain personnel carrying conveyors. Unless crash barriers are installed between the conveyor and vehicle track it is a sensible precaution to stop vehicles when persons are being carried on the conveyor. Control arrangements provided at some mines use traffic or other indicating lights which have an inbuilt timer to permit persons to complete a conveyor journey before the vehicle driver claims the road.

Appointment and training of personnel

15. Considerable variation in FSV driver training arrangements exist at mines. It is desirable that the initial driver training should take place in controlled surroundings, preferably on the surface, where an instructor can safely witness actions and give tuition before drivers proceed to the 'on-the-job' training. The current trend is for mines to stand alone in the training of drivers but if a central FSV training establishment is not available, it may be practicable for groups of mines to cooperate and share basic training facilities.

16. Supervisors and instructors should monitor a trainee driver's progress and an independent third party assessment by an examiner at the end of the training should secure consistency of standards. Periodic driver refresher training, to enable skill levels to be enhanced, bad practices to be eliminated, and safe practices to be updated are recommended. Familiarisation or refresher training is also necessary for transient contract drivers to acquaint them of the Managers Transport Rules and operating conditions at particular mines.

17. Medical examinations, including eye sight and hearing tests, are recommended prior to the appointment of drivers and periodically thereafter.

Photographs 4 & 5: Passenger carrying pod and purpose designed vehicle

Photograph 6: A typical well-designed FSV roadway with good clearances

18. It is the manager's duty to appoint persons (normally deputies) to be in charge of roadways in which FSVs operate. These appointees require special training as regards FSV operation and required roadway standards. Their general training and experience may not qualify them for these special duties. An FSV roadway daily report form designed for coal mines is desirable to aid reporting of floor conditions, clearances, obstructions, etc. Lengths of roadways should be zoned to assist the reporting of defects and the issuing of instructions for remedial work.

19. More comprehensive surveys should be conducted at periodic intervals by a competent person appointed by the manager.

Risks to pedestrians from FSVs

20. The risks presented to pedestrians in FSV roads may be obvious but are highlighted in Paragraphs 31 to 36. Experience has revealed that persons walk in roadways where FSVs operate for a variety of reasons; to supervise a trainee driver, to assist drivers in loading and offloading vehicles en route, acting as spotters to guide vehicles through places of limited clearance, etc. Systems of work should aim to eliminate the need for these practices and more widespread use of manriding vehicles, could also assist in reducing pedestrian activity.

Photograph 7: A purpose designed grading machine

Figure 1:
Unloaded flat bed vehicles

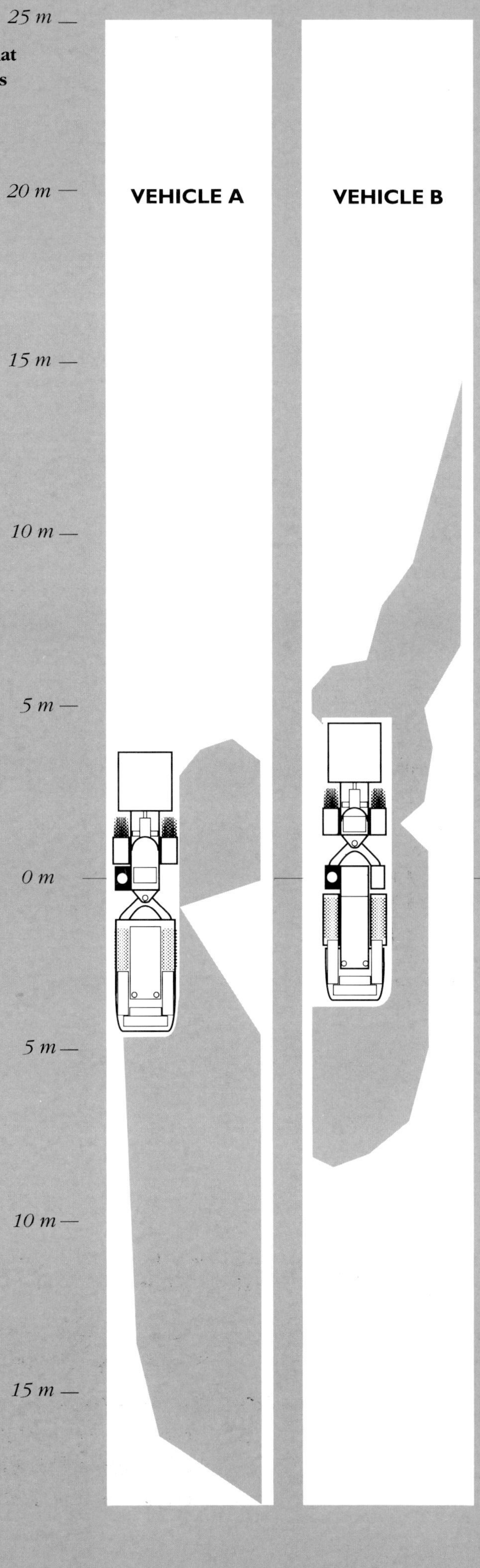

25 m
20 m — VEHICLE A   VEHICLE B
15 m
10 m
5 m
☐ Driver   0 m
5 m
10 m
15 m

Cab design and driver's vision

21. Driver vision is affected by a number of factors including the need for general low height vehicle design to suit mine roadways, the preference for centrally placed cabs to suit bi-directional travel and the consequential need for low cab seat location. The fitting and design of protective cabs and the design of any cab protective structure also have a significant influence on driver safety and vision. Driver's cabs, particularly in coal mines where headroom may be severely restricted, are expected to provide a roof or canopy over the driver's head to prevent his contact with the roof or other fixed overhead obstructions. It generally provides the reference point from which statutory vertical clearances are determined. The provision of a canopy may restrict the driver's vision by dictating that a fairly low driving position is adopted. This restriction is frequently further exacerbated by the presence of inadequate design in the cab supporting structure which may give rise to a trapping hazard, should the driver attempt to improve his vision by leaning out of the cab side. Serious consideration should be given to cab design having regard to the control position, available clearances and conditions of use. This may result in enhanced driver safety and vision with very minor changes to the overall design of the vehicle.

22. Surveys have been conducted to determine visibility limitations. Surveying instruments were used to plot blind areas where objects 1m and 1.8m high could not be seen by drivers, although they were allowed normal eye and head movement.

23. Figure 1 illustrates the variance in driver visibility for two makes of similarly sized flatbed vehicles when unloaded. The driver sits on the front chassis of vehicle **A** and on the rear chassis of vehicle **B**.

Objects less than 1m high not visible in shaded areas

A
Driver seated on front chassis

B
Driver seated on rear chassis

Figure 2: Loaded flat bed vehicles

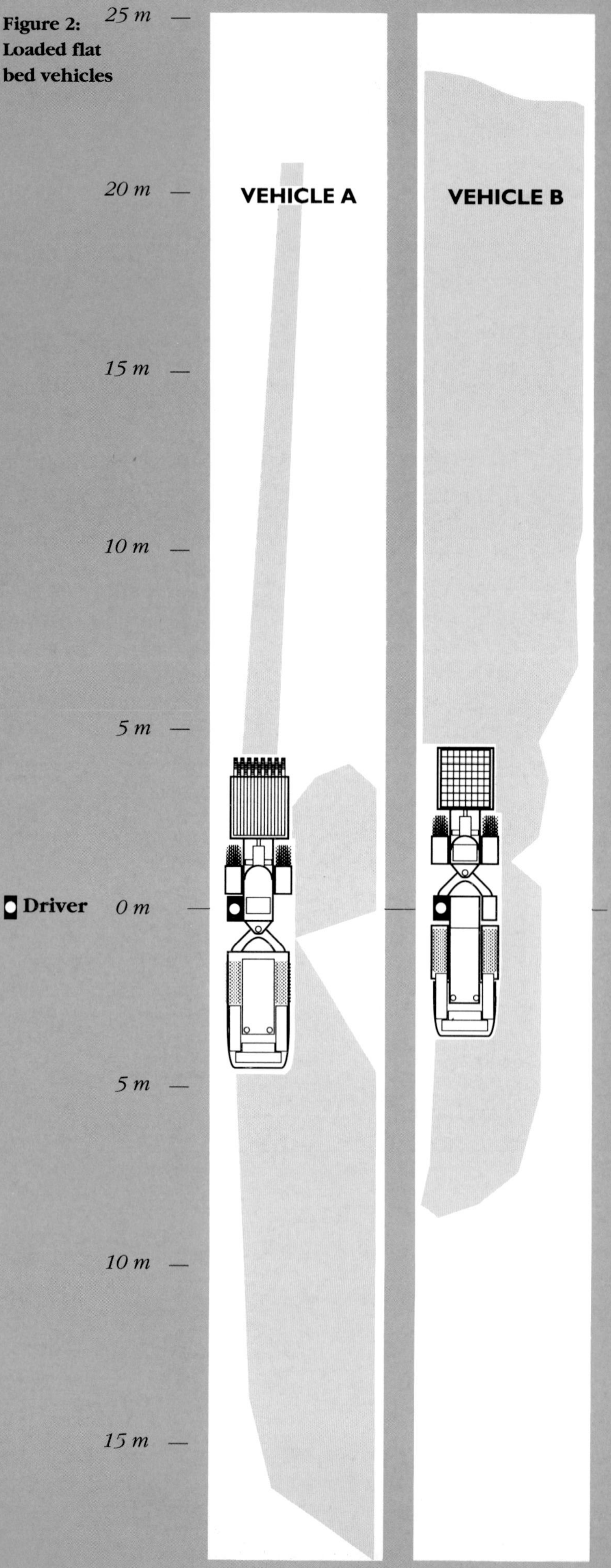

25 m
20 m — **VEHICLE A**   **VEHICLE B**
15 m
10 m
5 m
◻ Driver 0 m
5 m
10 m
15 m

24. Figure 2 shows the extended blind areas when vehicles are loaded with typical material loads. Vehicle **A** is loaded with an arch pack and vehicle **B** with wood chocks.

Objects less than 1m high not visible in shaded areas

Driver seated on front chassis

Driver seated on rear chassis

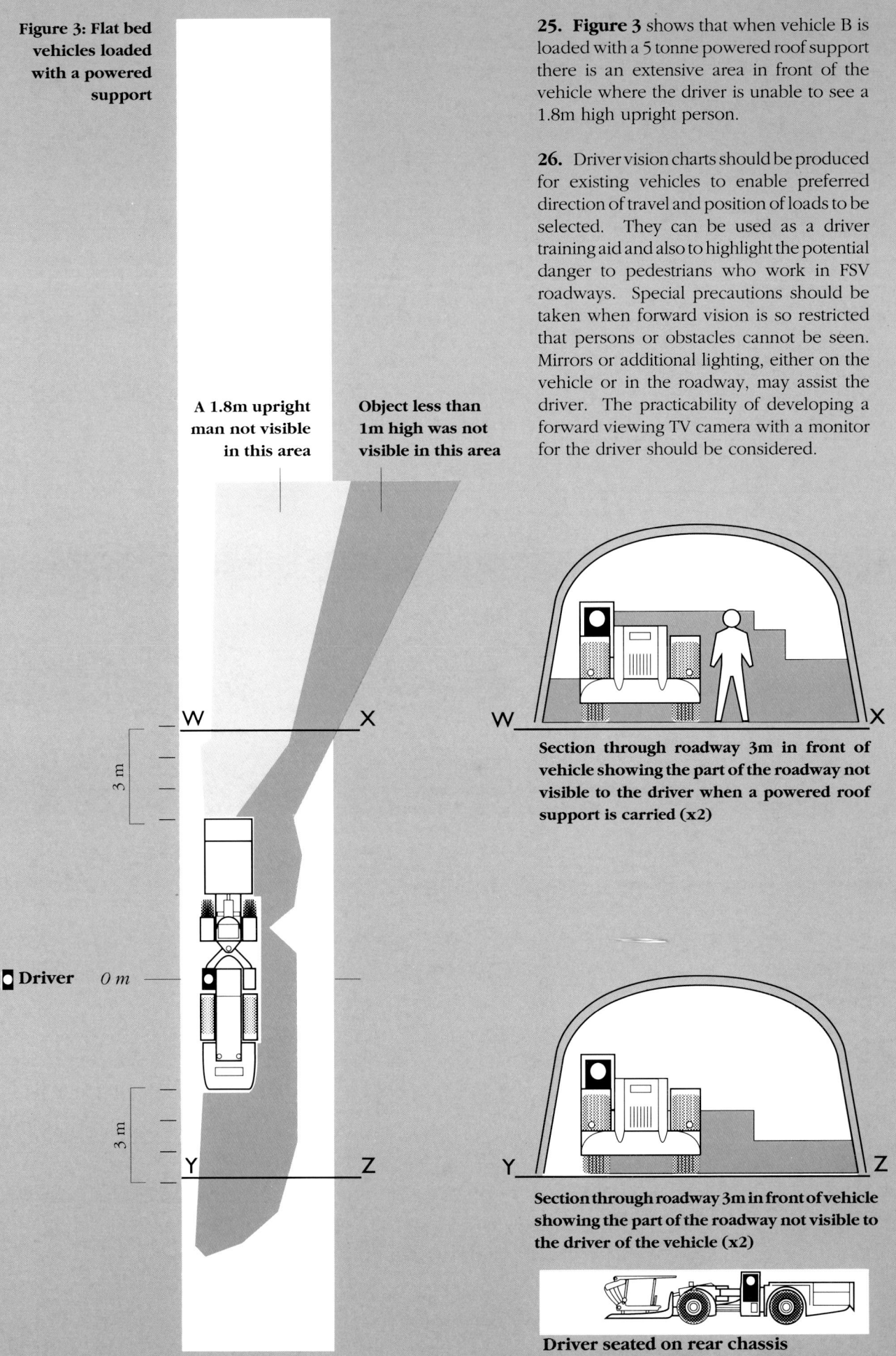

Figure 3: Flat bed vehicles loaded with a powered support

A 1.8m upright man not visible in this area

Object less than 1m high was not visible in this area

Section through roadway 3m in front of vehicle showing the part of the roadway not visible to the driver when a powered roof support is carried (x2)

Section through roadway 3m in front of vehicle showing the part of the roadway not visible to the driver of the vehicle (x2)

Driver seated on rear chassis

25. Figure 3 shows that when vehicle B is loaded with a 5 tonne powered roof support there is an extensive area in front of the vehicle where the driver is unable to see a 1.8m high upright person.

26. Driver vision charts should be produced for existing vehicles to enable preferred direction of travel and position of loads to be selected. They can be used as a driver training aid and also to highlight the potential danger to pedestrians who work in FSV roadways. Special precautions should be taken when forward vision is so restricted that persons or obstacles cannot be seen. Mirrors or additional lighting, either on the vehicle or in the roadway, may assist the driver. The practicability of developing a forward viewing TV camera with a monitor for the driver should be considered.

A REVIEW OF ACCIDENTS RESULTING FROM THE USE OF FSVs

Overall Accident Summary

27. A safety comparison of FSVs with other forms of underground transport is difficult to achieve because of the wide diversification of operating conditions. However, during the five-year period, April 1986 to March 1991, there were 11 fatal accidents and 282 major injury accidents involving rope haulage, five fatal and 74 major injury accidents involving locomotives and four fatal and 38 major injuries primarily attributed to the use of FSVs (HSE statistics).

28. **Table 1** lists FSV accident figures for each year of the study period. The unbracketed figures are accidents with some form of FSV involvement and figures in brackets represent those accidents where an FSV was considered to be the primary cause.

29. The principal hazards associated with the movement of FSVs are limited clearances and poor driver visibility, compounded by careless personal positioning and unsafe practices. Other hazards are associated with the loading or unloading of material, and with maintenance and servicing. Accident analysis by category is shown in **Table 2**.

TABLE 1

YEAR	FATAL	MAJOR INJURY	OVER 3 DAYS
1986 / 87	0	7 (5)	28
1987 / 88	0	9 (7)	49
1988 / 89	3 (3)	19 (12)	56
1989 / 90	0	6 (5)	48
1990 / 91	2 (1)	17 (9)	48
TOTALS	5 (4)	58 (38)	229

Note:
White figures: accidents with some FSV involvement (BCC Source)
Black figures: accidents with FSV as primary cause (HSE Source)

TABLE 2.

FSV accidents by category 1986-91

CATEGORY	FATAL	MAJOR INJURY	OVER 3 DAYS
ACCIDENTS DUE TO THE MOVEMENT OF FSVs	4	27	58
ACCIDENTS DURING LOAD TRANSFER	1	15	84
ACCIDENTS DURING MAINTENANCE	0	14	49
OTHERS	0	2	38
TOTALS	5	58	229

ACCIDENTS RESULTING FROM THE MOVEMENT OF VEHICLES

Summary

30. Fifty per cent of all FSV fatal and major injury accidents involve moving vehicles. These accidents are subdivided into those involving pedestrians, those resulting from a collision and miscellaneous.

Accidents to pedestrians

31. FSV operations underground pose a substantial risk to pedestrians unless adequate precautions are taken. In the period under review, two men were killed and five sustained major injury when run over by FSVs. Both fatal accidents were unwitnessed and in both instances the drivers were unaware that the accidents had occurred. Evidence suggested that those killed may have been sitting or lying down, resting with their cap lamps removed. Four of the major injury accidents also occurred to persons who were resting. The limitation of driver visibility is referred to in Paragraphs 21 to 26. The proximity of some accident sites to roadway junctions and turns, and the noise produced by auxiliary fans and other equipment masking the sound of the FSV approaching, are considered to have contributed to a number of accidents.

32. Eleven men sustained major injuries when they were struck or trapped by FSVs. Five occurred to men standing in the roadside to allow vehicles to pass. An overman directing operations to free a bogged down vehicle in a wet area was trapped against the roadside when it slewed unexpectedly. Six accidents were the result of persons being struck by shuttle car power cables. Many of the over 3 day accidents resulted from similar causes.

33. Supervisors and drivers too readily accept the presence of pedestrians in close proximity to vehicles. It is noteworthy that of the 18 fatal and major injury accidents five occurred to transfer point attendants, five to craftsmen and four to officials. The transient nature of the work of craftsmen and officials makes them particularly vulnerable and the need for and the timing of their movements should be examined.

TABLE 3.

Accidents resulting from the movement of vehicles

			FATAL	MAJOR INJURY	OVER 3 DAY
	ACCIDENTS TO PEDESTRIANS	RUN OVER	2	5	4
		STRUCK BY VEHICLE	0	6	7
		TRAPPED BY VEHICLE	0	5	7
	COLLISION ACCIDENTS	WITH ROADWAY SUPPORT	0	2	12
		WITH ROADWAY FITTING	0	2	6
		DRIVER TRAPPED	1	2	3
	OTHER ACCIDENTS	STRUCK BY TRAMP MATERIAL	0	2	5
		SECONDARY USE OF VEHICLE	0	3	4
		OTHER	1	0	10
TOTALS			4	27	58

Figure 8: A hazard zone at a roadway junction

36. Section 40 of the MQA 1954 requires refuge holes where vehicles run and Regulation 59(1) of the Coal and Other Mines (Shafts, Outlets and Roads) Regulations 1960 specifies the maximum intervals between them. Exceptions to these general requirements have been made in a few instances where FSVs have been introduced under Local Regulations.

Accidents resulting from collisions

37. Included in this group are accidents resulting from vehicles colliding with roadway sides and obstacles, and those where drivers are trapped by their vehicles.

38. An unusual fatal accident occurred when a high pressure water hydrant broke during attempts to seal it using a long steel bar inserted through the handwheel. The static head of water contained in the pipe range violently ejected the valve spindle which struck the deceased. Shortly before the accident, the valve had been damaged when struck by girders carried by an FSV. Another accident occurred when a vehicle cab struck a pressure reducing valve and the driver sustained serious injuries when leaving the cab to escape the water jet.

34. Any pedestrian working in or travelling along a road where FSVs run should be considered to be at risk. Wherever practicable, pedestrians should be kept clear of roads where FSVs are operating. Hazard or cautionary zones **(photograph 8)** should be established at places where it is unavoidable that vehicles and pedestrians will meet, eg. junctions, roadheads, transfer points and where persons have been deployed to repair or maintenance work. Manager's Transport Rules should specify the additional precautions to be taken when vehicles operate in these zones. Where persons have routinely to work in FSV roadways 'safe havens' should be provided wherever possible. One innovation has been to install pre-recorded talking messages at some hazard zones which can be activated by drivers and warn that an FSV is approaching.

39. An FSV driver was killed when trapped between a roof joist and the cab canopy. The cause of the accident is unclear, but it appears that the driver attempted to alight from the moving vehicle or was leaning out to improve his line of vision.

40. Vehicles colliding with roadway supports or fittings while negotiating turns or manoeuvring in material transfer stations caused three major injury accidents. No injuries have resulted from collision between vehicles.

35. In the context of visibility generally, and with the need for all persons below ground to be readily seen, especially where FSVs operate, the use by all persons below ground of reflective garments including general work wear, helmets, belts and cap lamp materials is recommended **(photographs 9a and 9b)**. Whilst add-on adhesive reflective strips are frequently used with the above-mentioned equipment, there is a need for the future manufacture of such equipment to incorporate reflective elements by design.

41. Vehicles colliding with roadway supports or fixtures caused 18 over 3 day accidents, 12 due to roadway supports being hit and six when pipes, ducting, etc. were struck. Three drivers were trapped when protruding outside their cabs. Collisions usually result in injury to drivers, but five pedestrians sustained injury when struck by dislodged supports or fitments. Cab collisions with arch supports injured six drivers.

Figure 9a & 9b: Photographs taken at 25m distance showing effective use of reflective markings on garments worn by a workman

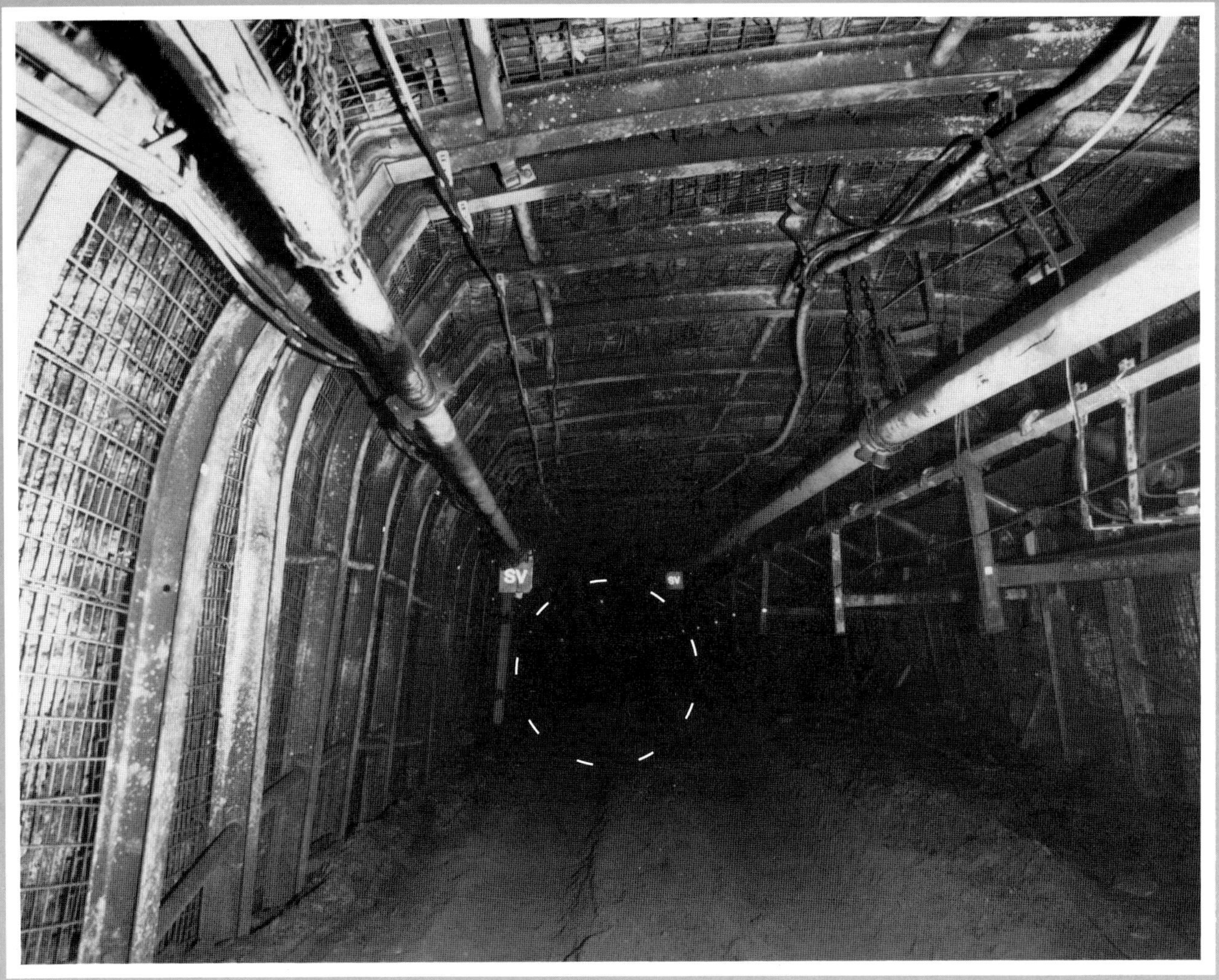
SV

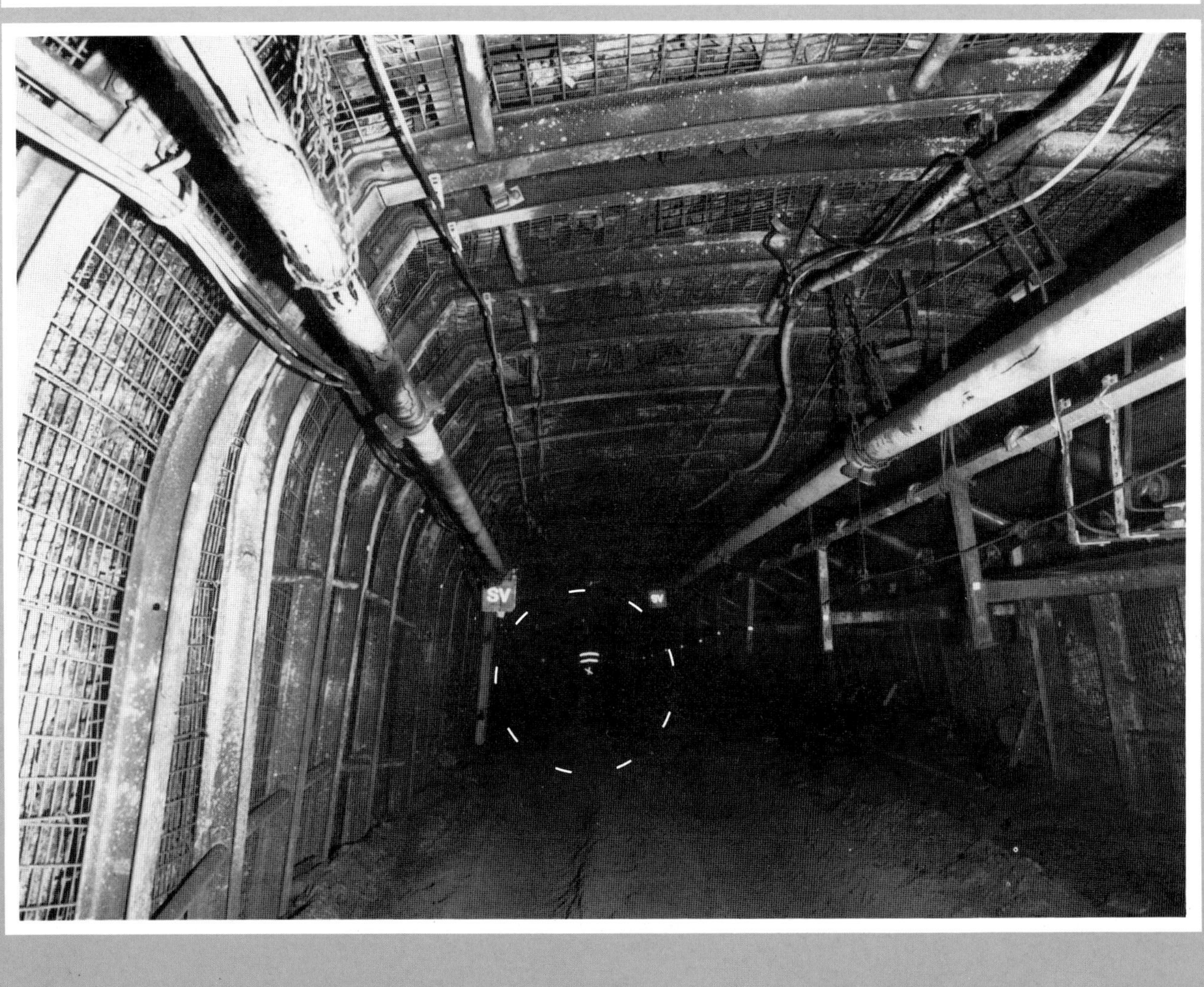
SV

42. The practice of driving FSVs with the cab side positioned close to the roadside provides maximum clearance on the drivers' 'blind side' but increases the risk of cab collisions and driver injury. Wet and rutted floor conditions feature in accidents where vehicles skid, tilt or bounce resulting in roadside contact. The often poor visibility from FSV cabs makes it difficult for drivers to see the floor and obstacles ahead. FSVs are lengthy vehicles and clearances can be difficult to judge when turning corners or manoeuvring and drivers are tempted to lean out of cabs in order to obtain a better view. Paragraphs 21 to 26 deal with driver visibility in some detail.

Clearances

43. The clearances specified in regulations and consents are minimum values and should not be regarded as initial design clearances particularly when roadway crush or floor heave is foreseeable. Employing pedestrians to guide drivers through roadways of reduced cross section to avoid conveyors or other equipment contact is potentially dangerous.

44. Clearances should be checked before any roadway is commissioned for FSV use and thereafter at intervals specified in the Manager's Transport Rules.

45. So far as possible, roadways should be dedicated to FSV use and should not include other equipment. Where other equipment is installed an evaluation of its safe positioning should be made. Vulnerable equipment should be identified and protected by crash barriers. Where height permits consideration should be given to the suspension of equipment above the operating level of the vehicles. Where necessary, effective use should be made of reflective markings strategically placed on equipment to serve as a guidance aid to drivers.

46. Statutory clearances apply equally to ventilation door frames and other obstacles as to the roof and sides.

47. Corners of junctions should be tapered to allow adequate clearances to be maintained and additional roadway width should be provided where FSV manoeuvring will occur.

Other accidents

48. Two major injury accidents resulted from tramp material on the floor being projected by vehicles, in one case injuring the driver, and in the other, a pedestrian. Four over 3 day accidents occurred to persons who were injured when struck by objects which slewed when run over. Good housekeeping, with an emphasis on tidiness, will eliminate such accidents.

49. The secondary use of vehicles for towing, placing or dragging resulted in three major injury and four over 3 day accidents to pedestrians. This highlights the need for adequate supervision of abnormal activities and for drivers to ensure pedestrians are well clear of vehicles before commencing operations.

TABLE 4.		FATAL	MAJOR INJURY	OVER 3DAY
Load transfer accidents	LOAD OUT OF CONTROL	1	7	29
	TRAPS AND STRAINS WHEN HANDLING LOADS	0	2	25
	STUMBLE, FALL, OR SLIP	0	1	14
	INJURED FITTING OR RELEASING LOAD BINDERS	0	3	14
	OTHERS	0	2	2
	TOTALS	1	15	84

FSV LOAD TRANSFER ACCIDENTS

Summary

50. Accidents occurring during the loading or unloading of materials, machinery or mineral accounted for 34% of all FSV related accidents and are categorised in **Table 4**.

Load out of control accidents

51. Accidents occur when loads slip or topple off FSVs during loading or unloading. One fatal accident, and several major injury accidents, occurred when heavy equipment was being handled. The majority of the over 3 day accidents involved bulk loads such as pipes, arches or girders.

52. One man was killed and two others narrowly escaped injury when a powered roof support toppled when being unloaded. As the FSV flat-bed was being lowered to the floor the support base fouled a lump of dirt causing the support to overturn trapping a workman. In another serious accident, a chock fitter was trapped against the roadside when a powered support, which had been lowered onto packing timbers, toppled over as the FSV withdrew.

53. Many inbye destinations have no permanent load handling facilities. FSVs are often unloaded by lowering the load to the floor and reversing away. This can only be done safely if all persons are well clear and the floor is reasonably smooth and level. Better load control would be achieved by using power operated hoists with pendant controls which allow the operator to be positioned a safe distance from the load. In relation to heavy handling tasks managers should determine when it is necessary to invoke a formal written system of work.

Strain and trapping injuries

54. Most of the accidents in this category resulted in strained back or trapped finger injuries sustained while manhandling bulk loads of girders, pipes, etc. Purpose designed tools to grip and handle such materials should be considered, but better packaging and more care when loading supplies would reduce the risks when offloading.

Stumbling, falling and slipping

55. Stumbling, falling and slipping accidents generally occurred when stepping on or off vehicle load platforms while manually transferring material loads. Slippery and untidy floors are also contributory factors. Monorail mechanical lifting and handling equipment, used for transferring loads, has the added advantage that loads can be positioned and stored away from the place where the vehicle stops so that clearances need not be affected by stacked materials.

Fitting and releasing loadbinders

56. Excessive effort is often used to apply loadbinders to ensure tightness and prevent loads shaking loose during transit. The number of accidents occurring while applying or releasing loadbinders suggest the system should be studied by engineers and alternative forms of restraint considered for particular loads. Webbing load binders on trial in some mines have ratchet type controllable tensioning and these devices can be fitted with chain intermediate sections if wear or chaffing is a problem.

Other accidents

57. Accidents occurred when previously stacked material was disturbed while vehicles were being offloaded. Load transfer stations can be made safer by stacking material in a specified location outside the offloading area.

Recommendations regarding transfer

58. Lifting and handling procedures which ensure safe positioning of persons should be established. Much of the heavy manual handling could be eliminated by greater use of monorails or other lifting equipment, especially where similar loads are regularly handled, and particularly at material transfer stations, face ends and headings. Consideration should be given to the use of gripping tools to lift arches, girders, pipes and rails to protect the hands and all persons employed on material handling should be trained in safe lifting techniques. Loading and unloading areas should be kept tidy and material should not be stored in these areas unless adequate space remains for loading and unloading to be done safely.

ACCIDENTS ASSOCIATED WITH FSV MAINTENANCE

Accident summary

59. Accidents which occurred during FSV maintenance, or in maintenance areas, are shown in **Table 5**.

Maintenance accidents

60. Accidents occurring during maintenance accounted for 22% of all FSV accidents. Scalding from cooling systems, chemical burns from cleaning fluids, contact with rotating engine fan blades, and injuries resulting from high pressure water hoses whipping out of control during cleaning down, are recurring types of accidents. Safety education of maintenance personnel, specifying the procedures to be followed and highlighting potential hazards, would help to reduce these accidents. Primary guarding of cooling fans is required to safeguard persons needing to carry out maintenance work with engine covers removed and the engine running.

61. Over a quarter of maintenance accidents resulted from persons slipping off vehicles, or from stumbling and falling in the garage area. A similar number resulted from manhandling items such as cabs, engine covers, and wheels. Discipline in replacing pit covers, tidiness, the provision of mechanical handling equipment, together with steps and raised platforms for working on the upper part of vehicles, would reduce such accidents.

Underground garages

62. Consents granted by the Mines Inspectorate for the use of FSVs require the same standard of housing stations, commonly known as garages, as those specified for diesel locomotives in the Locomotive Regulations. Garages are provided as a place for maintenance and for parking vehicles when not in use. The place designated for parking should be separate from, and should not necessitate persons travelling through, the maintenance area. Garage maintenance areas should be kept clear of vehicles other than those undergoing maintenance.

TABLE 5.

Maintenance accidents

		FATAL	MAJOR INJURY	OVER 3DAY
SERVICING OR FITTING	SCALDING	0	1	6
	CHEMICAL BURNS	0	0	5
	STRUCK BY ROTATING PART, E.G. FAN	0	3	2
	STRUCK BY HP WATER HOSE	0	3	0
	REMOVING PARTS	0	2	7
STUMBLING, FALLING AND SLIPPING	FALLING OFF VEHICLE	0	1	8
	IN SERVICE AREA	0	2	5
	FELL INTO PIT	0	0	3
MANHANDLING	COVERS AND CABS	0	1	7
	WHEELS	0	1	2
OTHERS	-	0	0	4
TOTALS		**0**	**14**	**49**

63. Two types of FSV maintenance garage are in general use. Central garages which are generally well designed and equipped, and inbye garages which, due to their temporary nature have more limited facilities for minor maintenance. Management should specify what tasks are to be carried out at inbye garages and ensure that all vehicles have available to them a garage with full maintenance facilities.

Safety awareness training

64. Safety awareness training, to highlight the potential for accidents in garages, should be given to persons who maintain FSVs. Supervisors should enforce tidiness and ensure that personal protective equipment is available and used.

OTHER ACCIDENTS INVOLVING FREE STEERED VEHICLES

Accident summary

65. Accidents not included in the three previous categories are summarised in **Table 6**.

Roadway floors

66. The formation of wheel ruts may indicate that road drainage should be improved or that localised areas require surfacing by concrete or some similar material. Trials in progress with a purpose made floor grading machine may prove to be a partial answer in roadways where floor evenness is difficult to maintain.

OTHER DANGERS

Underground fires

67. The restriction in the use of flammable materials in vehicle construction, the limitation on diesel engine and exhaust surface temperatures, the use of fire resistant fluids in hydraulic systems, flameproofing and electrical design and protection have all contributed to minimise the number of fires involving FSVs. Proper maintenance and good refuelling and battery charging arrangements have also helped to prevent such incidents.

68. Diesel engine FSVs may typically contain 20 litres of mineral oil and 125 litres of diesel fuel. It is customary in coal mines to fit manually activated fire suppression systems within the engine compartment. Some have fire detection systems which alarm to the driver.

69. Tyres are also combustible and the introduction at some mines of puncture proof polyurethane (PU) filled tyres resulted in fire tests being carried out in 1986 by HSE's Research and Laboratories Services Division at Buxton. These tests confirmed that temperatures and toxic gas concentrations were significantly higher for a PU filled tyre than a pneumatic one. The initial burning stages of these filled tyres are, however, slower and more time is available to extinguish a PU filled tyre before a serious hazard to life develops. Roadways in which FSVs operate should be provided with fire fighting hyrdrants.

TABLE 6.

	FATAL	MAJOR INJURY	OVER 3 DAY
PEDESTRIANS STUMBLING INTO WHEEL RUTS	0	2	13
DRIVER INJURED BOARDING OR ALIGHTING FSV	0	0	10
MISCELLANEOUS	0	0	15
TOTALS	0	2	38

70. In the five year study period, there were five FSV associated fires and no ignitions of gas by FSVs. Frictional heating of exposed disc brakes was the cause of all the fires which were extinguished by onboard portable extinguishers. Fires have primarily occurred on the high speed external brakes of shuttle cars. These highlight the need to keep brakes clean to allow unobstructed operation and to provide access for cooling ventilation. Modern designs of FSVs tend to have oil immersed disc brakes which run with more controlled temperatures.

Runaways

71. Investigations by Inspectors into a number of incidents of uncontrolled movement of FSVs has led to some design modifications. Many of the runaways occurred after the vehicle had been parked and the driver had left the cab. Parking brakes are normally applied by springs and released by hydraulic pressure controlled by a push/pull lever in the cab. Stored energy remains in accumulators in the hydraulic circuit after the engine has stopped, or electrical power has been switched off, and this is capable of releasing a park brake if the brake lever is inadvertently moved to the off position. The stored energy needs dissipating after parking to immobilise a vehicle effectively and this is generally achieved by a few movements of the hydraulic steering or by other appropriate means. Start up can also cause parking brakes to be suddenly released if the brake lever is left in the off position. Drivers should immobilise vehicles properly before leaving the cab and ensure that they are seated in the cab before attempting to start a vehicle. The need for drivers to leave their vehicles on gradients should be avoided if possible and moveable obstructions, eg. ventilation doors, should be power operated to aid this approach. The effect of a runaway would be significantly reduced if the good practice of angling vehicles into the roadside when parking was followed.

Overturning of FSVs

72. For the purpose of this report, FSV overturning means the vehicle overturning onto its side. In the five-year study period, three FSVs were reported to have overturned. However, instances when vehicles overbalance to rest against the roadside and are recoverable by the driver may pass unreported. Conveyor spillage, debris in the roadway and uneven floors contributed to the reported incidents. The high centre of gravity of certain loads, if positioned to one side of a vehicle platform, can cause overturning. Good roadway maintenance prevents obstacles accumulating or floor conditions deteriorating to cause excessive cross gradients.

73. When a pivot steered vehicle is turned onto a descending gradient, the centre of gravity of the vehicle and load moves to the inside of the turn to a position near the inside front wheel and an unstable situation can develop. The overturning risk is reduced if such roadways are designed with short level sections to allow vehicles to turn on the level rather than directly onto a steep gradient.

74. PU filled tyres have been used to allow vehicles to carry heavier loads, especially large powered supports. Stability is enhanced by the reduction in tyre deflection but this increases ground contact pressure and may cause greater floor damage. There is scope for research into PU tyre fill pressures and how this affects the maximum allowed load, tyre life and the comfort of ride for the driver. Fill pressures must be adequate to ensure tyre bead grip onto wheel rims as low pressure PU fill tyres have been known to rotate around the wheels.

75. Pneumatic tyres generally ensure a softer ride and should have ready means for air pressures to be checked and topped up.

Vehicle collisions

76. Paragraphs 37-42 discussed accidents where vehicles collided with roadway sides or obstacles, or where drivers were trapped by their vehicles. Collision damage to vehicles is a major and persistent problem and one of the primary reasons for vehicles being out of use. It is likely that there is considerable underreporting of collision events. Managers should instruct drivers to report all collisions to their supervising officials even when there has been no injury. These should be investigated to enable the problem to be assessed and appropriate remedial action taken to prevent a recurrence.

BRAKE TESTING

77. Brake testing methods used with FSVs vary considerably. Test aims should ensure that the inherently good brake design standards are maintained throughout. Any brake deterioration should be detected well before vehicle safety is affected. The Inspectorate have encouraged the development of instrumented static and dynamic brake testing to replace simple stopping tests for which the assessment is based simply on the opinion of drivers or mechanics. One static test system utilises an anchored rig with a hydraulic ram to pull a braked vehicle, pressure gauges indicating the force needed to overcome the brakes. Front and rear brakes on coal mine FSVs can normally be tested separately. Most vehicles have fourwheel drive with a transmission lock and wheel skid is not a problem when front and rear brakes are pull tested separately. Service, emergency and parking brakes can be tested in this way.

78. The SIMRET electronic retardometer, developed by HSE, can also be used to measure brake effort. At least one manufacturer is developing a SIMRET with a graphical printout of test results and it is anticipated that this will shortly be certified for use in coal mines. Trials are also being conducted with instruments which record speed and stopping distances.

79. Vehicle Consents and Regulations require weekly dynamic brake tests. In the short term, they may continue to be simple stopping distance tests which are best conducted with a known payload at a specified place with a measured gradient. Periodic instrument brake tests should become the norm in the not too distant future.

80. Parking brake tests should be able to demonstrate that the vehicle can be held when carrying its maximum allowed load on the steepest gradient with a sufficient margin of safety.

DIESEL VEHICLE EXHAUST EMISSIONS

81. The extent of diesel exhaust pollution is related to engine design. Engines suitable for use below ground may have been derated by limiting the fuel delivery to reduce carbon monoxide (CO) emission and the timing may have been retarded to control the production of oxides of nitrogen (NO_x). The specified engine settings should not be tampered with. Total exhaust gases emitted into the mine ventilation depends on the engine's work load, which is particularly high for load haul dump operations.

82. The undiluted exhaust of a diesel FSV in any type of mine must not contain more than 2000 parts per million (ppm) of CO or 1000ppm of NO_x. The production of sulphur dioxide depends on the sulphur content of the diesel fuel used. All fuel used below ground must comply with that specified in regulations or consents.

83. Statutory requirements for sampling undiluted exhaust gases specify that engines must be tested when idling with the vehicle stationary and when operating under maximum power. The stationary low idle test presents no difficulty, but full load exhaust sampling of a moving vehicle with no seat provided for the person taking the sample can be difficult and some mines have substituted this test by stationary high (full revs) engine idle sampling. This is not entirely satisfactory as it does not sample the exhaust in the most onerous operating conditions. Mines should review their exhaust sampling procedures and, if full load exhaust samples are not being taken, examine how this can best be achieved. It may be possible to use a portable battery powered onboard sampling pump which can be operated by the driver.

84. It has long been recognised that there is a possible carcinogenic risk from particulate matter in exhaust emissions. The difficulty in determining concentrations of diesel particulates is in differentiating between these particles and particles of coal dust of a similar size. The strict limitations in coal mines placed on diesel engine surface temperatures and exhaust levels may conflict with the use of devices which can be fitted to reduce particulate emissions. However, water bath conditioners provided on FSVs partly reduce particulate emissions. The most recent research into possible carcinogenicity caused by diesel engine exhaust emissions from FSVs in coal mines was carried out by the Institute of Occupational Medicine (IOM) and the results published in January 1989[*]. The research concluded that "provided vehicles are well maintained, careful consideration is given to ventilation requirements and good working practices are followed, there appears to be very little reason for not using FSVs for transporting men and materials on the grounds of exhaust emissions and associated risks to health."

85. The HSE, in conjunction with British Coal, has recently initiated a research programme which aims to develop techniques for measuring diesel particulate matter in coal mines. The effectiveness of particulate control systems, marketed for surface diesel engines, and their suitability for flame proofed underground engines, together with the possible beneficial effects of fuel additives, are among further matters being researched.

86. Good ventilation standards should always prevent flammable and noxious gases exceeding prescribed levels. Current consents identify two action levels. The first requires that if carbon monoxide (CO) or nitrogen dioxide (NO_2) in the general ventilation exceeds 50ppm or 3ppm by volume respectively, the manager must take steps to improve the ventilation to prevent those levels being exceeded, and the second requires diesel vehicles to be stopped if the respective levels exceed 100ppm or 5ppm.

BATTERY POWERED VEHICLES USED IN COAL MINES

87. Battery powered vehicles used in coal mines must be approved by HSE under the Electricity at Work Regulations 1989. A condition of HSE approval is that battery FSVs require the consent of the local HM Principal District Inspector before they operate in a particular part of a mine.

88. The Electricity at Work Regulations also require electricity to be cut off from any place below ground in a mine in the event of the flammable gas exceeding 1.25% by volume in the general body of air. Since

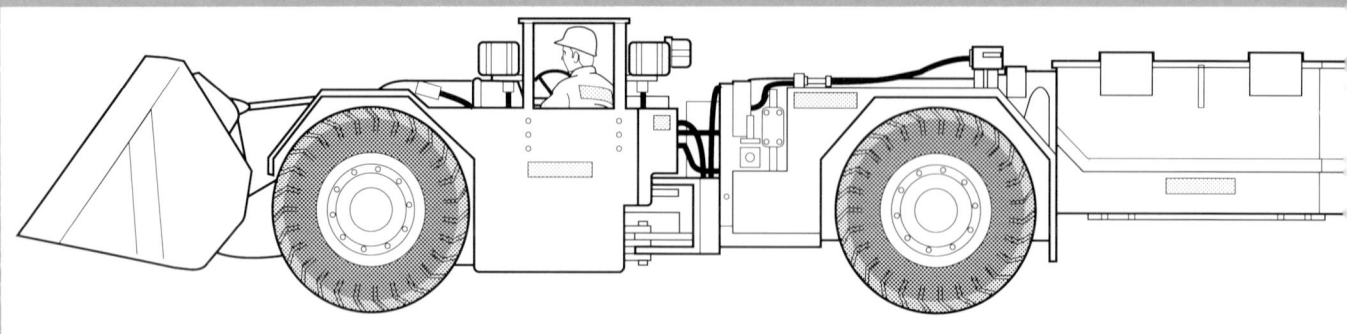

present technology is unable to remove the energy contained in a battery, other means have to be implemented to ensure safe operation. This was highlighted in the report into the explosion at Weetslade Colliery in 1951 (Cmd 9614) which killed five men and involved battery powered shuttle cars. The hazard can be reduced by removing battery vehicles to a safe place when predetermined levels of flammable gas are reached and well before the electricity cut off level. This level is normally set at 0.8% by volume of the general body air but should be related to the normal readings of flammable gas at that workplace.

---

[*]Further Environmental Aspects of the Use of Diesel Powered Equipment in Coal Mines - A Robertson, R P Garland, B Cherrie, J R D Nee - Report No. TM/89/01 - UDC 622.00.621.43.06

89. The report into the explosion at Eppleton Colliery in 1951 (Cmd 8503), in which nine men died, recommended that "no shuttle cars be used in any 'broken' workings (namely where the extraction of pillars in bord and pillar working is practised), nor, except with the written consent of an inspector, in any other working place in direct contact with, or about to hole through on, any goaf in a safety lamp mine". The recommendation was made because of the risk of sparking from protected but open ventilated battery containers used on the shuttle cars. These restrictions on the use of battery powered vehicles have since been extended to include areas liable to outbursts of flammable gas. Manager's Rules governing the use of battery FSVs usually depict on a plan those places where vehicles are not permitted to run.

90. Batteries should be designed to meet the requirements of the harmonised European Standard EN50019 "Increased Safety". Additionally, an isolator should be fitted to the outgoing terminals and another at the mid voltage point. With both isolators open, the risk of danger arising from electrical faults is reduced. Additionally, should abnormal conditions suddenly occur (eg. an outburst of flammable gas) the battery can be left with both isolators open and in a quiescent state until the potential danger is removed. Where batteries are to be stored below ground at places other than a recognised charging station they must always be left with both the main and midpoint isolators open.

91. Environmental monitoring in places where battery vehicles operate should primarily be concerned with the detection and warning of flammable gas at a level which would permit safe vehicle withdrawal. All auxiliary ventilated headings, and any other roadway or place where a manager perceives a significant risk from flammable gas, should be continually monitored. Strategically placed inroad monitors should give audible and visual alarms throughout the roadway when alarm levels of flammable gas have been detected. It is recommended that drivers carry portable automatic detectors so that they can be assured that the flammable gas content has not risen above the alarm level to the level at which electricity must be cut off, whilst vehicles are being withdrawn from gassy places. In auxiliary ventilated headings the flow of air should also be monitored and alarms given throughout the heading if the flow is significantly reduced, and vehicles should be withdrawn. Drivers should be prohibited from entering roadways where an alarm has been given.

92. In relation to the future design of both battery and diesel powered FSVs, the desirability of fitting onboard fire damp detection systems should be considered.

93. Due to the danger of igniting hydrogen gas, flame safety lamps and hot wire monitoring instruments should not be used within 10m of battery charging stations. Fire detection monitors may be a useful addition at battery storage and charging stations, particularly if the stations are unattended.

NOISE CONTROL

94. The principal noise sources of diesel powered FSVs are the engine and ancillaries, the cooling fan, hydraulics and transmission. Diesel engine noise results from combustion and mechanical movements, and indications are that engine noise is not related significantly to engine size but is affected by speed and load. Noise control is generally achieved by enclosing the engine compartment. Total enclosure is difficult to achieve and small open areas considerably reduce the affects of attenuation. Acoustic enclosure conflicts with engine cooling needs, and maintaining engine covers is a continuous task. The driver's exposure to noise can be reduced by fitting a barrier between the power pack and the driver and providing 'noise absorbent' cab roof linings.

95. Free steered vehicles approved by HSE since 1977 have been required to be designed to keep noise levels to the lowest practicable level and as far as is reasonably practicable below 90dB(A) at the driver's head during normal underground operations. The Noise at Work Regulations 1989 also apply. Noise surveys by Inspectors have found some drivers exposed to noise levels in the order of 96dB(A) indicating the need for more effective and reliable noise control measures.

CONCLUSIONS

96. The use of modern FSVs has revolutionised many areas of underground transport, not least the installation and salvaging of powered roof supports and their speedy transfer from one face to another. FSVs are here to stay, but it should not be assumed that they are always the most appropriate transport system; locomotives, rope haulages and cable powered crawler tracked transporters still have their place. FSVs eliminate many of the well known dangers of other methods of transport, eg. derailments and broken ropes, but they introduce new hazards, especially those associated with the absence of good forward vision for the drivers of bidirectional vehicles with centrally placed cabs. Mine managers should address these hazards when specifying FSV roadway design and in the preparation of Transport Rules. Such measures will go a considerable way to reducing FSV accidents, but the need for absolute self discipline by all who work with FSVs will remain of paramount importance.

SUMMARY OF RECOMMENDATIONS

97. Satisfactory clearances for the proposed use of FSVs should be designed into roadway drivages and these should take into account anticipated reduction in cross section due to strata movement. Additional clearances should be considered at bends, junctions, materials transfer sections, garages and other specified places. (paragraphs 10 to 12 and 43)

98. Unless crash barriers are installed between a manriding conveyor and the vehicle track it is a prudent precaution to stop vehicles operating when persons are being carried on the conveyor. (paragraph 14)

99. Supervisors and instructors should monitor a trainee driver's progress and an independent third party assessment by an examiner made at the end of the training. (paragraph 16)

100. Drivers should be given periodic refresher training. Any trained contract drivers, who may move from mine to mine, should be acquainted with the Managers Rules and operating conditions at particular mines before being authorised. (paragraph 16)

101. Medical examinations, including eye sight and hearing tests, are recommended prior to the appointment of drivers and periodically thereafter. (paragraph 17)

102. Persons appointed to be in charge of FSV roadways should be given special training in safe FSV operation and required roadway standards. (paragraph 18)

103. A daily report form should be designed to aid the reporting of roadway conditions relating, in particular, to the state of floors, clearances and obstructions. Roadways should be identified and zoned where necessary to assist the reporting of defects and the instruction of remedial work. (paragraph 18)

104. Comprehensive surveys of vehicle clearances, including those at ventilation doors and obstacles, should be made by competent persons appointed by the manager before a roadway is commissioned for FSV use, and at periodic intervals specified in the Manager's Transport Rules. (paragraphs 19 and 44 to 46)

105. Serious consideration should always be given to cab design having regard to the control position, available clearances and conditions of use. (paragraph 21)

106. Driver vision charts should be prepared for all vehicles to enable preferred direction of travel and position of load to be selected. (paragraph 26)

107. Special precautions should be taken when a driver's forward view is so restricted that persons or obstacles in close proximity cannot be seen. Mirrors either on the vehicle

or in the roadway and additional lighting should be provided where practicable. (paragraph 26)

108. The practicality of developing a forward viewing TV camera and monitor for the driver should be investigated. (paragraph 26)

109. Where possible, pedestrians should be kept out of roadways where FSVs operate. Those who are required to be present should be warned that they may not be seen by an approaching FSV driver and should be prohibited from sitting or lying in travelling roads. Where persons have routinely to work in FSV roadways 'safe havens' should be provided wherever possible. (paragraphs 31 and 34)

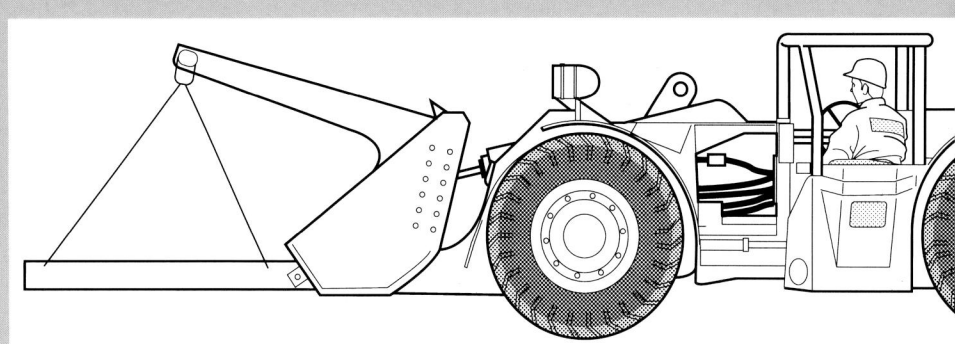

110. All persons below ground should wear reflective garments including general workwear, helmets, belts and cap lamp batteries and such equipment should be inherently reflective by design. (paragraph 35)

111. As far as possible, roadways should be dedicated to FSVs and not include other unnecessary equipment. An evaluation of the safe positioning of other equipment should be made and any which is vulnerable should be protected by crash barriers. The suspension of equipment above FSVs should be considered. (paragraph 45)

112. Effective use should be made of reflective markings strategically placed on equipment to serve as a guidance aid to drivers. (paragraph 45)

113. The abnormal use of FSVs for towing, dragging or positioning of equipment should be supervised by competent persons. (paragraph 49)

114. The practice of off loading FSVs by lowering the load to the floor and reversing away should only be done if the floor is reasonably smooth and level and all persons are well clear. (paragraphs 52 to 53)

115. For heavy handling tasks, managers should determine when it is necessary to invoke a formal written system of work. Power operated hoists with pendant controls to distance the operator from the load should be used more regularly. (paragraph 53)

116. The packaging of bulk loads of girders, pipes, etc. should be reviewed to reduce off loading dangers. The use of purpose designed tools to grip and handle such material should be considered as a means to prevent injuries from slipping and falling loads. (paragraph 54)

117. The present use of load binders should be studied by engineers and assessment made as to whether they are being used correctly or if they are always the most appropriate method of load restraint. Webbing load binders with ratchet type controllable tensioning should be considered. (paragraphs 56 and 58)

118. Engine cooling fans should have primary guarding. (paragraph 60)

119. Steps and raised platforms should be provided for maintenance staff when working on upper parts of vehicles. (paragraph 61)

120. Management should clarify which tasks are to be carried out in inbye garages where facilities are limited. All vehicles should have access to a garage with full maintenance provisions. (paragraphs 62 to 63)

121. FSV maintenance areas should be kept free of vehicles other than those being worked on. (paragraph 62)

122. Safety awareness training should be given to FSV maintenance staff to acquaint them of the types of accident which may occur in FSV garages. (paragraph 64)

123. Roadways in which FSVs operate should be provided with fire hydrants. (paragraph 69)

124. Drivers should immobilise vehicles before leaving the cab and be seated in the cab before attempting to start the vehicle. (paragraph 71)

125. The need for drivers to leave cabs regularly while vehicles are parked on gradients should be avoided. (paragraph 71)

126. After the engine of a parked vehicle has been stopped, the hydraulic energy stored in accumulators should be dissipated by the driver operating the hydraulic steering a few times or by other appropiate means. (paragraph 71)

127. All parked vehicles should be angled into the roadside to prevent runaways. (paragraph 71)

128. Excessive cross gradients caused by conveyor spillage, objects on the floor or uneven floors should be avoided. In order to minimise the risk of overturning it is desirable that vehicles turn on the level rather than directly on to gradients. (paragraphs 72 to 73)

129. Research into the effect of PU tyre fill pressures on the maximum allowed load, tyre life and driver comfort should be carried out. Fill pressure must be adequate to prevent PU fill tyres rotating around the wheels. (paragraph 74)

130. Pneumatic tyres should have ready means for air pressures to be checked and topped up. (paragraph 75)

131. Managers should instruct drivers to report all collisions to their supervising officials, even when there has been no injury, and these should be investigated as appropriate. (paragraph 76)

132. Brake testing should ensure that the inherent good design standards when new are maintained. Instrumented static and/or dynamic brake testing should be developed and become the norm in the not too distant future. In the short term, where testing is by measuring stopping distances, this should be conducted with a known payload on a measured gradient. (paragraphs 77 to 80)

133. Approved engine settings must not be tampered with. Diesel exhaust sampling must be carried out with the engine at both low idle and under full load. Mines should review their sampling procedures to ensure compliance with statutory requirements. (paragraphs 81 to 86)

134. Batteries should be constructed to EN50019 and have midpoint isolators which should be opened together with the main isolator whenever a battery is to be stored at a place other than a charging station.(paragraph 90)

135. Environmental monitoring in places where battery vehicles operate should be concerned with the detection of flammable gas at a level which permits safe vehicle withdrawal. Drivers should carry portable firedamp detectors on their vehicles. Auxiliary ventilated headings or other places where a manager perceives a significant risk of flammable gas, should be constantly monitored. The flow of air in auxiliary ventilated headings should be constantly monitored. (paragraph 91)

136. In relation to the future design of both battery and diesel powered FSVs, the desirability of fitting onboard firedamp detection systems should be considered. (paragraph 92)

137. Flame safety lamps and hot wire monitoring instruments should not be used in close proximity to battery charging stations. (paragraph 93)

138. Noise measurements should be made to ensure FSV drivers are not exposed to levels in excess of 90dB(A) during normal vehicle operations at mines. (paragraph 94 to 95)

GLOSSARY

FSV — Free Steered Vehicle

HSE — Health and Safety Executive

HMIM — HM Inspectorate of Mines

MQA — Mines and Quarries Act 1954

CONSENT — A document issued under Section 83 MQA, permitting the use of diesel powered vehicles below ground in a mine and describing conditions for their use. Special regulations have been issued as an alternative in the past.

APPROVED — Vehicles of a type approved by HSE for use in mines where flammable gas is or might be a hazard.

TM12 — The HSE Testing Memorandum TM12, dated 1977, which describes procedures and requirements for the testing and approval of diesel and storage battery vehicles for use in mines.

BCC — British Coal Corporation

QDS — Quick Detachable System for quick fitment and removal of attachments such as buckets, flat beds, manriding pods, etc.

LHD — Load Haul Dump vehicle; a vehicle capable of self loading, hauling and discharging.

SHUTTLE CAR — A vehicle designed to receive, haul and discharge but not to self load.

GLOSSARY